This handbook is dedicated to the memory

of

Diti Hengchaovanich

Geotechnical Engineer

of

Thailand.

He pioneered the use of vetiver on a large scale for highway stabilization, and for many years was a very valuable contributor to The Vetiver Network International. Diti will be remembered with gratefulness by many.

1st Edition 2008
Published by The Vetiver Network International

Cover by Lily Grimshaw

PREFACE

THE VETIVER SYSTEM
FOR SLOPE STABILIZATION

AN ENGINEER'S HANDBOOK

The Vetiver System (VS) is dependent on the use of a very unique tropical plant, vetiver grass, *Vetiveria zizanioides* – recently reclassified as *Chrysopogon zizanioides*. The plant can be grown over a very wide range of climatic and soil conditions, and if planted correctly can be used virtually anywhere under tropical, semi-tropical, and Mediterranean climates. It has characteristics that in totality are unique to a single species. When vetiver grass is grown in the form of a narrow self-sustaining hedgerow it exhibits special characteristics that are essential to many of the different applications that comprise the Vetiver System.

Vetiver grass can be used for applications that will protect river basins and watersheds against environmental damage, particularly from point source factors relating to: 1. sediment flows (often associated with agriculture and infrastructure), and 2. toxic chemical flows resulting from excess nutrients, heavy metals and pesticides in leachate from agriculture and other industries. Both are closely linked.

This handbook is a modified extraction from *Vetiver Systems Applications - A Technical Reference Manual* (2008) by Paul Truong, Tran Tan Van, and Elise Pinners, and focuses on the protection of infrastructure and for disaster mitigation by applying the Vetiver System to slope stabilization. It draws on ongoing vetiver work in Vietnam and elsewhere in the world. Its technical recommendations and observations are based on real life situations, problems and solutions. The handbook is primarily for engineers and others with res ponsibility for the construction and protection of infrastructure.

Dick Grimshaw
Founder and Chairman of The Vetiver Network International.

i

FORWARD

Based on the review of the huge volume of Vetiver System research and application, the authors considered that it was time to compile a new publication to replace the first World Bank published handbook (1987), Vetiver Grass - A Hedge Against Erosion (commonly known as the Green Book), prepared by John Greenfield. This handbook is one of three, and focuses on the use of the Vetiver System for infrastructure protection through its application for slope stabilization.

The handbook includes the most up to date R&D results and numerous examples of highly successful results from around the world and particularly from Vietnam, where an intensive country wide vetiver program has been introduced since 2000. The main aim of this handbook is to introduce VS to planners, design and construction engineers and other potential users involved with infrastructure at all levels, who often are unaware of the effectiveness of the Vetiver System for bio-engineering applications.

In addition to the information in this handbook there are many articles and research papers relating to the use of the Vetiver System for slope stabilization on the Vetiver Network's website at: www.vetiver.org.

Details about the authors, and acknowledgments of those who contributed to this handbook can be found in the master manual *Vetiver Systems Applications - A Technical Reference Manual* (2008). It is suffice to say that we deeply acknowledge and appreciate all those involved in this handbook production.

The principle author of this handbook is Tran Tan Van, Vice-Director of the Vietnam Institute of Geosciences and Mineral Resources in Vietnam and Coordinator of The Vietnam Vetiver Network.

Paul Truong, Tran Tan Van and Elise Pinners.
The authors.

THE VETIVER SYSTEM FOR SLOPE STABILIZATION

AN ENGINEER'S HANDBOOK

PART 1
VETIVER GRASS - THE PLANT

CONTENTS

1. INTRODUCTION

The Vetiver System (VS), which is based on the application of vetiver grass (*Vetiveria zizanioides* L Nash, now reclassified as *Chrysopogon zizanioides* L Roberty), was first introduced by the World Bank for soil and water conservation in India in the mid 1980s. While this application still plays a vital role in agricultural land management, R&D conducted in the last 20 years has clearly demonstrated that, due to vetiver grass' extraordinary characteristics, VS also has important application as a bioengineering technique for steep slope stabilization, wastewater disposal, phyto-remediation of contaminated land and water, and other environmental protection purposes.

What does the Vetiver System do and how does it work?
VS is a very simple, practical, inexpensive, low maintenance and very effective means of soil and water conservation, sediment control, land stabilizations and rehabilitation, and phyto-remediation. Being vegetative it is also environmental friendly. When planted in single

rows vetiver plants will form a hedge which is very effective in slowing and spreading run off water, reducing soil erosion, conserving soil moisture and trapping sediment and farm chemicals on site. Although many hedges can do this, vetiver grass, due to its extraordinary and unique morphological and physiological characteristics described below can do it better than all other systems tested. In addition, the extremely deep and massively thick root system of vetiver binds the soil and at the same time makes it very difficult for it to be dislodged under high velocity water flows. This very deep and fast growing root system also makes vetiver very drought tolerant and highly suitable for steep slope stabilization.

The Extension Workers Manual, or the Little Green Book
Complementing this handbook is the slim green extension workers pocket book first published be the World Bank in 1987 and referred to on page ii as Vetiver Grass - A Hedge Against Erosion, or more commonly known the "little green book" by John Greenfield. This handbook is far more technical in its description of the Vetiver System and is aimed at engineers, technicians, academics, planners and Government officials and land developers.

2. SPECIAL CHARACTERISTICS OF VETIVER GRASS

2.1 Morphological characteristics:
- Vetiver grass does not have stolons or rhizomes. Its massive finely structured root system that can grow very fast, in some applications rooting depth can reach 3-4m in the first year. This deep root system makes vetiver plant extremely drought tolerant and difficult to dislodge by strong current.
- Stiff and erect stems, which can stand up to relatively deep water flow - photo 1.
- Highly resistance to pests, diseases and fire - photo 2.
- A dense hedge is formed when planted close together acting as a very effective sediment filter and water spreader.
- New shoots develop from the underground crown making vetiver resistant to fire, frosts, traffic and heavy grazing pressure.
- New roots grow from nodes when buried by trapped

2

sediment. Vetiver will continue to grow up with the deposited silt eventually forming terraces, if trapped sediment is not removed.

Photo 1: Erect and stiff stems form a dense hedge when planted close together.

2.2 Physiological characteristics

- Tolerance to extreme climatic variation such as prolonged drought, flood, submergence and extreme temperature from -15°C to +55°C.
- Ability to regrow very quickly after being affected by drought, frosts, salinity and adverse conditions after the weather improves or soil ameliorants added.
- Tolerance to wide range of soil pH from 3.3 to 12.5 without soil amendment.
- High level of tolerance to herbicides and pesticides.
- Highly efficient in absorbing dissolved nutrients such as N and P and heavy metals in polluted water.
- Highly tolerant to growing medium high in acidity, alkalinity, salinity, sodicity and magnesium.
- Highly tolerant to Al, Mn and heavy metals such as As, Cd, Cr, Ni, Pb, Hg, Se and Zn in the soils.

2.3 Ecological characteristics

Although vetiver is very tolerant to some extreme soil and climatic conditions mentioned above, as typical tropical grass, it is intolerant to shading. Shading will reduce its growth and in extreme cases, may even eliminate vetiver in the long term. Therefore vetiver grows best

**Photo 2: Upper: Vetiver grass surviving forest fire;
lower: two months after the fire.**

in the open and weed free environment, weed control may be needed during establishment phase. On erodible or unstable ground vetiver first reduces erosion, stabilizes the erodible ground (particularly steep slopes), then because of nutrient and moisture conservation, improves its micro-environment so other volunteered or sown plants can establish later. Because of these characteristics vetiver can be considered as a nurse plant on disturbed lands.

Photo 3: On coastal sand dunes in Quang Bình (upper) and saline soil in Gò Công Province (lower).

Photo 4: On extreme acid sulfate soil in Tân An (upper) and alkaline and sodic soil in Ninh Thun (lower).

2.4 Cold weather tolerance of vetiver grass

Although vetiver is a tropical grass, it can survive and thrive under extremely cold conditions. Under frosty weather its top growth dies back or becomes dormant and 'purple' in colour under frost conditions but its underground growing points survived. In Australia, vetiver growth was not affected by severe frost at −14°C and it survived for a short period at −22°C (-8°F) in northern China. In Georgia (USA), vetiver survived in soil temperature of -10°C but not at −15°C. Recent research showed that 25°C was optimal soil temperature for root growth, but vetiver roots continued to grow at 13°C. Although very little shoot growth occurred at the soil temperature range of 15°C (day) and 13°C root growth continued at the rate of 12.6cm/day, indicating that vetiver grass was not dormant at this temperature and extrapolation suggested

6

that root dormancy occurred at about 5°C (Fig.1).

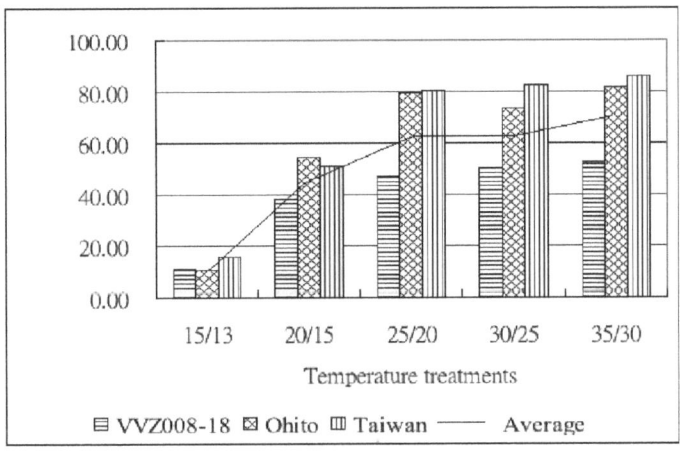

Figure 1: The effect of soil temperature on the root growth of vetiver.

2.5 Summary adaptability range

Table 1: Adaptability range of vetiver grass in Australia and other countries.

Condition characteristic	Australia	Other Countries
Adverse Soil Conditions		
Acidity (pH)	3.3-9.5	4.2-12.5 (high level soluble Al)
Salinity (50% yield reduction)	17.5 mScm^{-1}	
Salinity (survived)	47.5 mScm^{-1}	
Aluminium level (Al Sat. %)	Between 68% - 87%	
Manganese level	> 578 mgkg^{-1}	
Sodicity	48% (exchange Na)	
Magnesicity	2400 mgkg^{-1} (Mg)	

continued on next page ....

Condition characteristic	Australia	Other Countries
Fertilizer		
vetiver can be established on very infertile soil due to its strong association with mycorrhiza	N and P (300 kg/ha DAP)	N and P, farm manure
Heavy Metals		
Arsenic (As)	100 - 250 mgkg^{-1}	
Cadmium (Cd)	20 mgkg^{-1}	
Copper (Cu)	35 - 50 mgkg^{-1}	
Chromium (Cr)	200 - 600 mgkg^{-1}	
Nickel (Ni)	50 - 100 mgkg^{-1}	
Mercury (Hg)	> 6 mgkg^{-1}	
Lead (Pb)	> 1500 mgkg^{-1}	
Selenium (Se)	> 74 mgkg^{-1}	
Zinc (Zn)	>750 mgkg^{-1}	
Location	15^{0}S to 37^{0}S	41^{0}N - 38^{0}S
Climate		
Annual Rainfall (mm)	450 - 4000	250 - 5000
Frost (ground temp.)	-11^{0}C	-22^{0}C
Heat wave	45^{0}C	55^{0}C
Drought (no effective rain)	15 months	
Palatability	Dairy cows, cattle, horse, rabbits, sheep, kangaroo	Cows, cattle, goats, sheep, pigs, carp
Nutritional Value	N = 1.1 %	Crude protein 3.3%
	P = 0.17%	Crude fat 0.4%
	K = 2.2%	Crude fibre 7.1%

Genotypes: VVZ008-18, Ohito, and Taiwan, the latter two are basically the same as Sunshine. Temperature treatments: day 15°C /night 13°C (PC: YW Wang).

2.6 Genetic characteristics
Three vetiver species are used for environmental protection purposes.

2.6.1 *Vetiveria zizanioides* reclassified as *Chrysopogon zizanioides*
There are two species of vetiver originating in the Indian subcontinent: *Chrysopogon zizanioides* and *Chrysopogon lawsonii*. *Chrysopogon zizanioides* has many different accessions. Generally those from south India have been cultivated and have large and strong root systems. These accessions tend towards polyploidy and show high levels of sterility and are not considered invasive. The north Indian accessions, common to the Gangetic and Indus basins, are wild and have weaker root systems. These accessions are diploids and are known to be weedy, though not necessarily invasive. These north Indian accessions are NOT recommended under the Vetiver System. It should also be noted that most of the research into different vetiver applications and field experience have involved the south Indian cultivars that are closely related (same genotype) as Monto and Sunshine. DNA studies confirm that about 60% of *Chrysopogon zizanioides* used for bio-engineering and phytoremediation in tropical and subtropical countries are of the Monto/Sunshine genotype.

2.6.2 *Chrysopogon nemoralis*
This native vetiver species are wide spread in the highlands of Thailand, Laos, and Vietnam and most likely in Cambodia and Myanmar as well. It is being widely used in Thailand for thatching purpose. This species is not sterile, the main differences between *C. nemoralis* and *C. zizanioides*, are that the latter is much taller and has thicker and stiff stems, *C. zizanioides* has a much thicker and deeper root system and its leaves are broader and has a light green area along the mid ribs, as shown on the photos below - photos 5-8.

9

Photo 5: Vetiver leaves, upper: *C. zizanioides*, lower: *C. nemoralis*.

Photo 6: Difference between *C. zizanioides* (upper) and *C. nemoralis* roots (lower).

Photo 7: Vetiver shoots: upper - *C. nemoralis*, lower - *C. zizanioides*.

Photo 8: Vetiver roots after being grown in soil (top left and right), and after being grown suspended in water (lower).

Although *C. nemoralis* is not as effective as *C. zizanioides*, farmers have also recognized the usefulness of *C. nemoralis* in soil conservation; they have used it in the Central Highlands as well as in some coastal provinces of Central Vietnam such as Quang Ngai to stabilize dikes in rice fields, - photo 9.

Photo 9: *C. nemoralis* on a rice field bund in Quang Ngai (upper), and wild in Central Highlands (lower).

2.6.3 Chrysopogon nigritana
This species is native to Southern and West Africa, its application is mainly restricted to the sub continent, and as it produces viable seeds its application should be restricted to their home lands - photo 10.

13

2.7 Weed potential

Vetiver grass cultivars derived from south Indian accessions are non-aggressive; they produce neither stolons nor rhizomes and have to be established vegetatively by root (crown) subdivisions. It is imperative that any plants used for bioengineering purposes will not become a weed in the local environment; therefore sterile vetiver cultivars

Photo 10: *Chrysopogon nigritana* in Mali, West Africa.

(such as Monto, Sunshine, Karnataka, Fiji and Madupatty) from south Indian accessions are ideal for this application. In Fiji, where vetiver grass was introduced for thatching more than 100 years ago, it has been widely used for soil and water conservation purposes in the sugar industry for over 50 years without showing any signs of invasiveness. Vetiver grass can be destroyed easily either by spraying with glyphosate

(Roundup) or by cutting off the plant below the crown.

3. CONCLUSION

Due to *C. nemoralis* low growth forms and most importantly very short root system it is not suitable for steep slope stabilization works. In addition, no research has been conducted on its wastewater disposal and treatment, and phyto-remediation capacities, it is recommended that only non fertile cultivars of *C. zizanioides* be used for applications listed in this manual.

4. REFERENCES

Adams, R.P., Dafforn, M.R. (1997). DNA fingerprints (RAPDs) of the pantropical grass, *Vetiveria zizanioides* L, reveal a single clone, "Sunshine," is widely utilised for erosion control. Special Paper, The Vetiver Network, Leesburg Va, USA.

Adams, R.P., M. Zhong, Y. Turuspekov, M.R. Dafforn, and J.F.Veldkamp. 1998. DNA fingerprinting reveals clonal nature of *Vetiveria zizanioides* (L.) Nash, Gramineae and sources of potential new germplasm. Molecular Ecology 7:813-818.

Greenfield, J.C. (1989). Vetiver Grass: The ideal plant for vegetative soil and moisture conservation. ASTAG - The World Bank, Washington DC, USA.

National Research Council. 1993. Vetiver Grass: A Thin Green Line Against Erosion. Washington, D.C.: National Academy Press. 171 pp.

Purseglove, J.W. 1972. Tropical Crops: Monocotyledons 1. , New York: John Wiley & Sons.

Truong, P.N. (1999). Vetiver Grass Technology for land stabilisation, erosion and sediment control in the Asia Pacific region. Proc. First Asia Pacific Conference on Ground and Water Bioengineering for Erosion Control and Slope Stabilisation. Manila, Philippines, April 1999.

Veldkamp. J.F. 1999. A revision of Chrysopogon Trin. including *Vetiveria Bory* (Poaceae) in Thailand and Melanesia with notes on some other species from Africa and Australia. Austrobaileya 5: 503-533.

PART - 2
THE VETIVER SYSTEM
FOR
SLOPE STABILIZATION

CONTENTS

1. TYPES OF NATURAL DISASTERS THAT CAN BE REDUCED BY USING THE VETIVER SYSTEM (VS)

Besides soil erosion, the Vetiver System (VS) can reduce or even eliminate many types of natural disasters, including landslides, mudslides, road batter instability, and erosion (river banks, canals, coastlines, dikes, and earth-dam batters).

When heavy rains saturate rocks and soils, landslides and debris-flows occur in many mountainous areas of Vietnam. Representative examples are the catastrophic landslides, debris flows and flash flooding in the Muong Lay district, Dien Bien province (1996), and the landslide on the Hai Van Pass (1999) that disrupted North-South traffic for more than two weeks and cost more than $1 million USD to remedy. Vietnam's largest landslides, those larger than one million cubic meters (among them Thiet Dinh Lake, Hoai Nhon district, Binh Dinh province, in An Nghiêp and An Linh communes, Tuy An district, and Phu Yen province), caused loss of life as well as property damage.

River bank and coastal erosion, and dike failures happen continually throughout Vietnam. Typical examples include: river bank erosion in Phu Tho, Hanoi, and in several central Vietnam provinces (including Thua Thien Hue, Quang Nam, Quang Ngai and Binh Dinh); coastal erosion in Hai Hau district, Nam Dinh province, and; riverbank and coastal erosion in the Mekong Delta. Although these events and flooding/storm disasters usually occur during the rainy season, sometimes riverbank erosion takes place during the dry season, when water drops to its lowest level. This happened in Hau Vien village, Cam Lo district, in Quang Tri province.

Landslides are more common in areas where human activities play a decisive role. Almost 20 percent or 200 km (124 miles) of more than 1000 km (621 miles) of the Ha Tinh - Kon Tum section of the Ho Chi Minh Highway is highly susceptible to landslide or slope instability, mainly because of poor road construction practices and an underlying failure to understand the unfavourable geological conditions. Recent landslides in the towns of Yen Bai, Lao Cai, and Bac Kan followed municipal decisions to expand housing by allowing cutting at increased slope gradients.

Major earthquakes have also generated landslides in Vietnam, including the 1983 slide in Tuan Giao district, and the 2001 slide along the route from Dien Bien town to Lai Chau district.

From a strictly economic point of view, the cost of remediating these problems is high and the State budget for such works is never sufficient. For example, river bank revetment usually costs between US $200,000-300,000 /km, sometimes running as high as US $700,000-$1 million /km. The Tan Chau embankment in the Mekong Delta is an extreme case that cost nearly US $7 million /km. River bank protection in Quang Binh province alone is estimated to require an expenditure of more than US $20 million ; the annual budget is only US $300,000 .

Construction of sea dikes usually costs between US $700,000-$1 million /km, but more expensive sections can cost upwards of US $2.5 million /km, and are not uncommon. After storm No. 7 in September 2005 washed away many improved dike sections, some dike managers concluded that even sections engineered to withstand storms up to the 9th level are too weak, and began to seriously consider constructing sea dikes capable of withstanding storms of up to the 12th level that would cost between US $7-$10 million /km.

Budget constraints always exist, which confines rigid structural protection measures to the most acute sections, never to the full length of the river bank or coastline. This band aid approach compounds the problems.

Each of these events represents a type of slope failure or mass wasting, reflecting the down slope movement of rock debris and soil in response to gravitational stresses. This movement can be very slow, almost imperceptible, or devastatingly rapid and apparent within minutes. Since many factors influence whether natural disasters will occur, we should understand the causes as well as some basic principles of slope stabilisation. This information will allow us to effectively employ VS bioengineering methods to reduce their impact.

2. GENERAL PRINCIPLES OF SLOPE STABILITY AND SLOPE STABILISATION

2.1 Slope profile
Some slopes are gradually curved, and others are extremely steep. The profile of a naturally-eroded slope depends primarily on its rock/soil type, the soil's natural angle of repose, and the climate. For slip resistant rock/soil, especially in arid regions, chemical weathering is slow compared to physical weathering. The crest of the slope is slightly convex to angular, the cliff face is nearly vertical, and a debris slope is present at a 30-35° angle of repose, the maximum angle at which loose material of a specific soil type is stable.

Non-resistant rock/soil, especially in humid regions, weathers rapidly and erodes easily. The resulting slope contains a thick soil cover. Its crest is convex, and its base is concave.

2.2 Slope stability
2.2.1 Upland natural slope, cut slope, road batter etc.
The stability of such slopes is based on the interplay between two types of forces, driving forces and resisting forces. Driving forces promote down slope movement of material, while resisting forces deter movement. When driving forces overcome resisting forces, these slopes become unstable.

19

2.2.2 River bank, coastal erosion and instability of water retaining structures

Some hydraulic engineers may argue that bank erosion and unstable water retaining structures should be treated separately from other types of slope failure because their respective loads are different. In our opinion, however, both are subject to the same interaction between "driving forces" and "resisting forces". Failure results when the former overcomes the latter.

However, erosion of banks and the instability of water retaining structures are slightly more complicated; they result from interactions between hydraulic forces acting at the bed and toe and gravitational forces affecting the in-situ bank material. Failure occurs when erosion of the bank toe and the channel bed adjacent to the bank have increased the height and angle of the bank to the point that gravitational forces exceed the shear strength of the bank material. After failure, failed bank material may be delivered directly to the flow and deposited as bed material, dispersed as wash load, or deposited along the toe of the bank either as intact block, or as smaller, dispersed aggregates.

Fluvial controlled processes of bank retreat are essentially twofold. Fluvial shear erosion of bank materials results in progressive incremental bank retreat. Additionally, a rise in bank height due to near-bank bed degradation or an increase in bank steepness due to fluvial erosion of the lower bank may act alone or together to decrease the stability of the bank with respect to mass failure. Depending on the constraints of its material properties and the geometry of its profile, a bank may fail as the result of any one of several possible mechanisms, including planar, rotational, and cantilever type failures.

Non-fluvial controlled mechanisms of bank retreat include the effects of wave wash, trampling, and piping - and sapping-type failures, associated with stratified banks and adverse groundwater conditions.

2.2.3 Driving forces

Although gravity is the main driving force, it cannot act alone. Slope angle, angle of repose of specific soil, climate, slope material, and especially water, contribute to its effect:

- Failure occurs far more frequently on steep slopes than on gentle slopes.
- Water plays a key role in producing slope failure especially at the toe of the slope:
  - In the form of rivers and wave action, water erodes the base of slopes, removing support, which increases driving forces.
  - Water also increases the driving force by loading, that is, filling previously empty pore spaces and fractures, which adds to the total mass subjected to gravitational force.
  - The presence of water results in pore water pressure that reduces the shear strength of the slope material. Importantly, abrupt changes (dramatic increases and decreases) in pore water pressure may play the decisive role in slope failure.
  - Water's interaction with surface rock and soil (chemical weathering) slowly weakens slope material, and reduces its shear strength. This interaction reduces resisting forces.

2.2.4 Resisting forces

The main resisting force is the material's shear strength, a function of cohesion (the ability of particles to attract and hold each other together) and internal friction (friction between grains within a material) that opposes driving forces. The ratio of resisting forces to driving forces is the safety factor (SF). If SF >1 the slope is stable. Otherwise, it is unstable. Usually a SF of 1.2-1.3 is marginally acceptable. Depending on the importance of the slope and the potential losses associated with its failure, a higher SF should be ensured. In short, slope stability is a function of: rock/soil type and its strength, slope geometry (height, angle), climate, vegetation and time. Each of these factors may play a significant role in controlling driving or resisting forces.

21

2.3 Types of slope failure

Depending on the type of movement and the nature of the material involved, different types of slope failure may result:

Table 1: Types of slope failure

Type of movement		Material involved	
		Rock	*Soil*
Falls		Rock fall	Soil fall
Slides	Rotational	Rock slump block	Soil slump blocks
	Translational	Rock slide	debris slide
Flows	Slow	Rock creep	Soil creep
			saturated & unconsolidated material
			earth flow
			mudflow (up to 30% water)
	Fast		debris flow
			debris avalanche
Complex	Combination of two or more types of movement		

In rock, usually falls and translational slides (involving one or more planes of weakness) will occur. Since soil is more homogenous and lacks a visible plane of weakness, rotational slides or flows occur. In general, mass wasting involves more than one type of movement, for example, upper slump and lower flow, or upper soil slide and lower rock slide.

2.4 Human impact on slope failure

Landslides are natural occurring phenomena known as geological erosion. Landslides or slope failures occur whether people are there or not! However, human land use practices play a major role in slope processes. The combination of uncontrollable natural events (earthquakes, heavy rainstorms, etc.) and artificially altered land (slope excavation, deforestation, urbanisation, etc.) can create disastrous slope failures.

2.5 Mitigation of slope failure

Minimizing slope failure requires three steps: identification of potentially unstable areas; prevention of slope failure, and; implementation of corrective measures following slope failure. A thorough understanding of geological conditions is critically important to decide the best mitigation practice.

2.5.1 Identification

Trained technicians identify prospective slope failure by studying aerial photographs to locate previous landslide or slope failure sites, and conducting field investigations of potentially unstable slopes. Potential mass-wasting areas can be identified by steep slopes, bedding planes inclined toward valley floors, hummocky topography (irregular, lumpy-looking surfaces covered by younger trees), water seepage, and areas where landslides have previously occurred. This information is used to generate a hazard map showing the landslide-prone unstable areas.

2.5.2 Prevention

Preventing landslides and slope instability is much more cost effective than correction. Prevention methods include controlling drainage, reducing slope angle and slope height, and installing vegetative cover, retaining wall, rock bolt, or shotcrete (finely-aggregated concrete, with admixture for fast solidifying, applied by a powerful pump). These supportive methods must be correctly and appropriately applied by first ensuring that the slope is internally and structurally stable. This requires a good understanding of local geological conditions.

2.5.3 Correction

Some landslides can be corrected by installing a drainage system to reduce water pressure in the slope, and prevent further movement. Slope instability problems bordering roads or other important places typically require costly treatment. Done timely and properly, surface and subsurface drainage would be very effective. However, since such maintenance is usually deferred or neglected entirely, much more rigorous and expensive corrective measures become necessary.

In Vietnam, rigid structural protection methods (concrete or rock riprap bank revetment, groins, retaining walls, etc.) are commonly used to stabilize slopes and riverbanks and to control coastal erosion. Nevertheless, despite their continuous use for decades, slopes continue to fail, erosion worsens, maintenance costs increase. So what are the main weaknesses of these measures? From a strictly economic point of view, rigid measures are very expensive, and state or municipal budgets for such projects are never sufficient. A technical and environmental analysis raises the following concerns:

- Mining of the rock/concrete occurs elsewhere, where it undoubtedly wreaks environmental havoc.
- Localized rigid structural devices do not absorb flow/wave energy. Since rigid structures cannot follow the local settlement, they cause strong gradients. Strong gradients generate additional turbulence, which creates more erosion. Moreover, since the devices are localized, they frequently end abruptly; they do not transit gradually and smoothly to the natural bank. Thus, they simply transfer erosion to another place, to the opposite side or downstream, which aggravates the disaster, rather than reducing it for the river as a whole. Examples of these abound in several Central Vietnam provinces.
- Structural, rigid measures introduce considerable amounts of stone, sand, cement into the river system, displacing and disposing large volumes of bank soil into the river. As the river becomes silted up, its dynamics change, its bed rises, and flood and bank erosion problems increase. This problem is particularly grave in Vietnam where workers throw waste soil directly into the river as they re-shape the bank. Often they dump stone directly into the river to stabilize the toe of unstable bank, or try to lay rock pieces on the riverbed, which reduces the flow depth (channel) considerably. When the embankments ultimately fail, scraps of rock baskets, groins, etc. remain scattered in the water causing man-made aggradation of the river bed.
- Rigid structures are unnatural and are incompatible with

the soft ground of eroding or erodible soils. As the ground is consolidated and/or eroded and washed away, it undercuts and undermines the upper rigid layer. Examples include the right bank immediately downstream of the Thach Nham Weir (Quang Ngai province) that cracked and collapsed. Engineers who replace concrete plates with rock riprap with or without concrete frames leave unsolved the problem of subsurface erosion. Along the Hai Hau sea dike, the whole section of rock riprap collapsed as the foundation soil underneath was washed away.

- Rigid structures only temporarily reduce erosion. They cannot help stabilize the bank when big landslides with deep failure surface.
- Concrete or rock retaining walls are probably the most common engineering method employed to stabilize road batters in Vietnam. Most of these walls are passive, simply waiting for the slopes to fail. When the slopes do fail, the walls also fail, as seen in many areas along the Ho Chi Minh Highway. These structures are also destroyed by earthquakes.

Although rigid structures like rock embankments are obviously unsuitable for certain applications, such as sand dune stabilisation, they are still being built, as can be observed along the new road in central Vietnam.

2.6 Vegetative slope stabilisation

Vegetation has been used as a natural bioengineering tool to reclaim land, control erosion and stabilize slopes for centuries, and its popularity has increased markedly in the last decades. This is partly due to the fact that more information about vegetation is now available to engineers, and also partly due to the cost-effectiveness and environment-friendliness of this "soft" engineering approach.

Under the impact of the several factors presented above a slope will become unstable due to: (a) surface erosion or 'sheet erosion'; and (b) internal structural weaknesses. Sheet erosion when not controlled often leads to rill and gully erosion that, over time, will destabilize the slope; structural weakness will ultimately cause mass movement or landslip. Since sheet erosion can also cause slope failure, slope surface protection should be considered as important as other structural reinforcements but its importance is often over looked. Protecting the slope surface is an effective, economical, and essential preventive measure. In many cases, applying some preventive measures will ensure continued slope stability, and always cost much less than corrective measures.

The vegetative cover provided by grass seeding, hydro-seeding or hydro-mulching normally is quite effective against sheet erosion and small rill erosion, and deep-rooted plants such as trees and shrubs can provide some structural reinforcement for the ground. However, on newly-constructed slopes, the surface layer is often not well consolidated, so even well-vegetated slopes cannot prevent rill and gully erosion. Deep-rooted trees grow slowly and are often difficult to establish in such hostile territory. In these cases, engineers often rue the inefficiency of the vegetative cover and install structural reinforcement soon after construction. In short, traditional slope surface protection provided by local grasses and trees cannot, in many cases, ensure the needed stability.

2.6.1 Pros, cons and limitations of planting vegetation on slope.
Table 2: General physical effects of vegetation on slope stabilization.

Effect	Physical Characteristics
Beneficial	
Root reinforcement, soil arching, buttressing, anchorage, arresting the roll of loose boulders by trees	Root aeration, distribution and morphology; Tensile strength of roots; Spacing, diameter and embedment of trees, thickness and inclination of yielding strata; Shear strength properties of soils
Depletion of soil moisture and increase of soil suction by root uptake and transpiration	Moisture content of soil; Level of ground water; Pore pressure/soil suction

Interception of rainfall by foliage, including evaporative losses.	Net rainfall on slope
Increase in the hydraulic resistance in irrigation and drainage canals.	Manning's coefficient
Adverse	
Root wedging of near-surface rocks and boulders and uprooting in typhoons.	Root area ration, distribution and morphology
Surcharging the slope by large (heavy) trees (sometimes beneficial depending on actual situations).	Mean weight of vegetation
Wind loading.	Design wind speed for required return period; mean mature tree height for groups of trees
Maintaining infiltration capacity	Variation of moisture content of soil with depth

Table 3: Slope angle limitations on establishment of vegetation.

Slope angle (degrees)	Vegetation type	
	Grass	**Shrubs/Trees**
0 - 30	Low in difficulty; routine planting techniques may be used	Low in difficulty; routine planting techniques may be used
30 - 45	Increasingly difficult for sprigging or turfing; routine application for hydro seeding	Increasingly difficult to plant
> 45	Special consideration required	Planting must generally be on benches

2.6.2 Vegetative slope stabilisation in Vietnam

To a lesser extent, softer, vegetative solutions have also been employed in Vietnam. The most popular bioengineering method to control riverbank erosion is probably the planting of bamboo (which is the worst measure you can take. Once bamboo clumps washout in a flood and go down river they can take out bridges or anything they get caught up in. They have such high tensile strength they do not break up). To control coastal erosion, mangrove, casuarinas, wild pineapple, and nipa palm are also employed. However, these plants have some major deficiencies, for example:

- Growing in clumps, bamboo which is shallow rooted does not close as a hedgerow. Therefore floodwater concentrates at the gaps between clumps, which increases its destructive power and causes more erosion.
- Bamboo is top heavy. Its shallow (1-1.5 m deep) bunch root system does not balance the high, heavy canopy. Therefore, clumps of bamboo add stress to a river bank, without contributing to its stability.
- Frequently the bunch root system of bamboo destabilizes the soil beneath it, encouraging erosion and creating the conditions for larger landslides. Several Central Vietnam provinces display examples of bank failure following installation of extensive bamboo strips.
- Mangrove trees, where they can grow, form a solid buffer that reduces wave power, which, in turn, reduces coastal erosion. However, establishing mangrove is difficult and slow as mice eat its seedling. Typically, of the hundreds of hectares planted, only a small percentage survives to become forest. This has been reported recently in Ha Tinh province.
- Casuarinas trees have long been planted on thousands of hectares of sand dunes in Central Vietnam. Wild pineapple is also planted along banks of rivers, streams and other channels, and along the contour lines of dune slopes. Although they reduce wind power and minimize sand storm, these plants cannot stem sand flow because they have shallow root systems and do not form closed hedgerows. Despite planting casuarinas and wild pineapple trees atop the sand dikes along

flow channels in Quang Binh province, sand fingers continue to invade arable land. Moreover, both plants are sensitive to climate; casuarinas seedlings barely survive sporadic but extreme cold winters (less than -15°C/5°F), and wild pineapple cannot survive North Vietnam's blistering summers.

Fortunately, vetiver grows quickly, becomes established under hostile conditions, and its very deep and extensive root system provides structural strength in a relatively short period of time. Thus, vetiver can be a suitable alternative to traditional vegetation, provided that the following application techniques are learned and followed carefully.

3. SLOPE STABILISATION USING VETIVER SYSTEM

3.1 Characteristics of vetiver suitable for slope stabilisation
Vetiver's unique attributes have been researched, tested, and developed throughout the tropical world, thus ensuring that vetiver is really a very effective bioengineering tool:

- Although technically a grass, vetiver plants used in land stabilisation applications behave more like fast-growing trees or shrubs. Vetiver roots are, per unit area, stronger and deeper than tree roots.
- Vetiver's extremely deep and massive finely structured root system can extend down to two to three meters (six to nine feet) in the first year. On fill slope, many experiments show that this grass can reach 3.6m (12 feet) in 12 months. (Note that vetiver certainly does not penetrate deeply into the groundwater table. Therefore at sites with a high groundwater level, its root system may not extend as long as in drier soil). Vetiver's extensive, and thick root system binds the soil which makes it very difficult to dislodge, and extremely tolerant to drought.
- As strong or stronger than those of many hardwood species, vetiver roots have very high tensile strength that has been proven positive for root reinforcement in steep slopes.
- These roots have a mean tested tensile strength of about 75 Mega Pascal (MPa), which is equivalent to 1/6 of mild steel reinforcement and a shear strength increment of 39% at a

depth of 0.5m (1.5 feet).

- Vetiver roots can penetrate a compacted soil profile such as hardpan and blocky clay pan common in tropical soils, providing a good anchor for fill and topsoil.
- When planted closely together, vetiver plants form dense hedges that reduce flow velocity, spread and divert runoff water, and create a very effective filter that controls erosion. The hedges slow down the flow and spreads it out, allowing more time for water to soak into the ground.
- Acting as a very effective filter, vetiver hedges help reduce the turbidity of surface run-off. Since new roots develop from nodes when buried by trapped sediment, vetiver continues to rise with the new ground level. Terraces form at the face of

Photo 1: Vetiver forms a thick and effective bio-filter both above (upper) and below ground (lower).

30

the hedges, this sediment should never be removed. The fertile sediment typically contains seeds of local plants, which facilitates their re-establishment.

- Vetiver tolerates extreme climatic and environmental variation, including prolonged drought, flooding and submergence, and temperature extremes ranging from -14°C to 55°C (7° F to 131°F) (Truong et al, 1996).
- This grass re-grows very quickly following drought, frost, salt and other adverse soil conditions when the adverse effects are removed.
- Vetiver displays a high level of tolerance to soil acidity, salinity, sodicity and acid sulfate conditions (Le van Du and Truong, 2003).

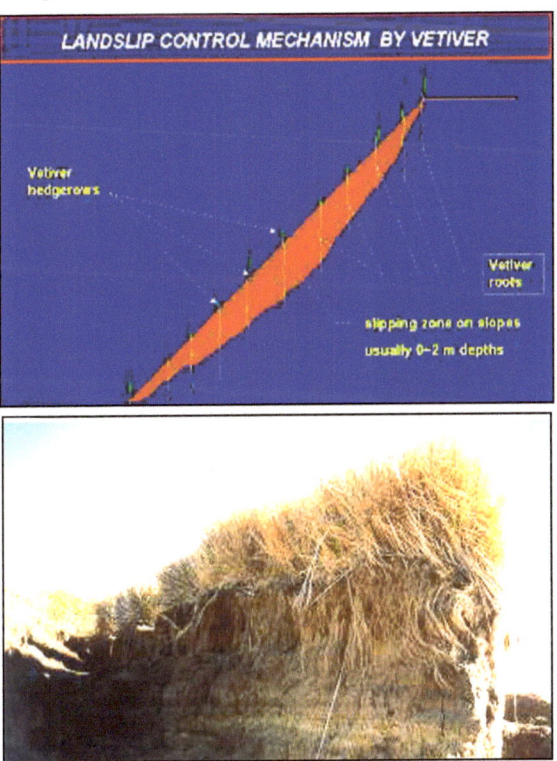

Figure 1: Upper: principles of slope stabilisation by vetiver; lower: vetiver roots reinforcing this dam wall kept it from being washed away by flood.

31

Vetiver is very effective when planted closely in rows on the contour of slopes. Contour lines of vetiver can stabilize natural slopes, cut slopes and filled embankments. Its deep, rigorous root system helps stabilize the slopes structurally while its shoots disperse surface run-off, reduce erosion, and trap sediments to facilitate the growth of native species. Hengchaovanich (1998) also observed that vetiver can grow vertically on slopes steeper than 150% (~56°). Its fast growth and remarkable reinforcement make it a better candidate for slope stabilisation than other plants. Another less obvious characteristic that sets it apart from other tree roots is its power of penetration. Its strength and vigour enable it to penetrate difficult soil, hardpan, and rocky layers with weak spots. It can even punch through asphalt concrete pavement. The same author characterizes vetiver roots as living soil nails or 2-3m (6-9 feet) dowels commonly used in 'hard approach' slope stabilisation work. Combined with its ability to become quickly established in difficult soil conditions, these characteristics make vetiver more suitable for slope stabilisation than other plants.

3.2 Special characteristics of vetiver suitable for water disaster mitigation

To reduce the impact of water related disasters such as flood, river bank and coastal erosion, dam and dike instability, vetiver is planted in rows either parallel to or across the water flow or wave direction. Its additional unique characteristics are very useful:

- Given its extraordinary root depth and strength, mature vetiver is extremely resistant to washouts from high velocity flow. Vetiver planted in north Queensland (Australia) has withstood flow velocity higher than 3.5m/sec (10'/sec) in river under flood conditions and, in southern Queensland, up to 5m/sec (15'/sec) in a flooded drainage channel.
- Under shallow or low velocity flow, the erect and stiff stems of vetiver act as a barrier that reduces flow velocity (i.e. increase hydraulic resistance) and traps eroded sediment. In fact, it can maintain its erect stance in a flow as deep as 0.6-0.8m (24-31").
- Vetiver leaves will bow under deep and high velocity flow, providing extra protection to surface soil while reducing flow velocity.

- When planted on water-retaining structures such as dams or dikes, vetiver hedgerows help reduce the flow velocity, decrease wave run-up (lap-erosion), over-topping, and ultimately the volume of water that flows into the area protected by these structures. These hedgerows also help reduce so-called retrogressive erosion that often occurs when the water flow or wave retreats after it rises over water-retaining structures.
- As a wetland plant, vetiver withstands prolonged submergence. Chinese research shows that vetiver can survive longer than two months under clear water.

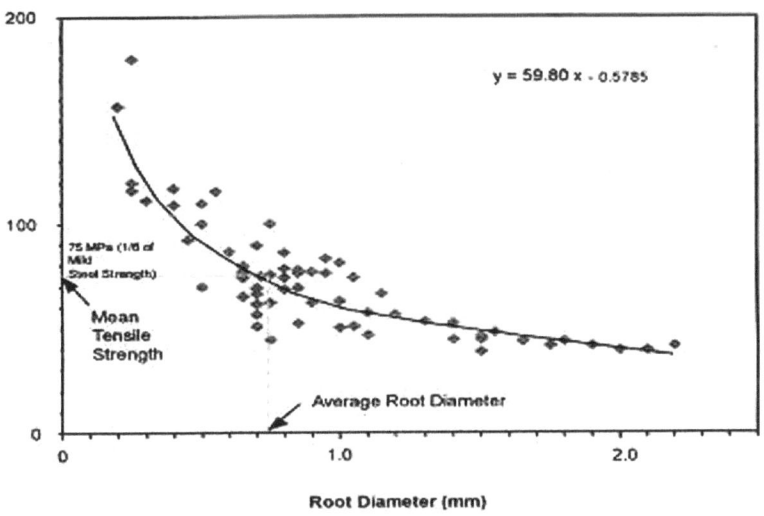

Figure 2: Root diameter distribution

3.3 Tensile and shear strength of vetiver roots

Hengchaovanich and Nilaweera (1996) show that the tensile strength of vetiver roots increases with the reduction in root diameter, implying that stronger, fine roots provide greater resistance than thicker roots. The tensile strength of vetiver roots varies between 40-180 MPa in the range of root diameter between 0.2-2.2 mm (.008-.08"). The mean design tensile strength is about 75 MPa at 0.7-0.8 mm (.03") root diameter, which is the most common size of vetiver roots, and equivalent to approximately one sixth of mild steel. Therefore, vetiver roots are as strong or even stronger than those of many hardwood species that have been proven positive for slope reinforcement - figure 2 and table 4.

Table 4: Tensile strength of some plant roots.

Botanical name	Common name	Tensile strength (MPa)
Salix spp	Willow	9-36
Populus spp	Poplars	5-38
Alnus spp	Alders	4-74
Pseudotsuga spp	Douglas fir	19-61
Acer sacharinum	Silver maple	15-30
Tsuga heterophylia	Western hemlock	27
Vaccinum spp	Huckleberry	16
Hordeum vulgare	Barley	15-31
	Grass, Forbs	2-20
	Moss	2-7kPa
Chrysopogon zizanioides	**Vetiver grass**	**40-120 (average 75)**

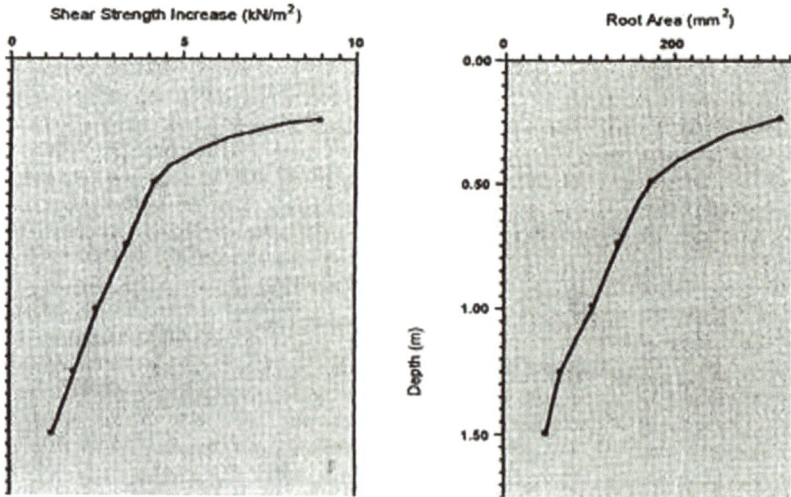

Figure 3: Shear strength of vetiver roots

In a soil block shear test, Hengchaovanich and Nilaweera (1996) also found that root penetration of a two-year-old vetiver hedge with 15cm (6") plant spacing can increase.

In a soil block shear test, Hengchaovanich and Nilaweera (1996) also found that root penetration of a two-year-old Vetiver hedge with 15cm (6") plant spacing can increase the shear strength of soil in adjacent 50 cm (20") wide strip by 90% at 0.25 m (10") depth. The increase was 39% at 0.50 m (1.5') depth and gradually reduced

to 12.5% at one meter (3') depth. Moreover, vetiver's dense and massive root system offers better shear strength increase per unit fibre concentration (6-10 kPa/kg of root per cubic meter of soil) compared to 3.2-3.7 kPa/kg for tree roots (Fig.3). The authors explained that when a plant root penetrates across a potential shear surface in a soil profile, the distortion of the shear zone develops tension in the root; the component of this tension tangential to shear zone directly resists shear, while the normal component increases the confining pressure on the shear plane.

Table 5: Diameter and tensile root strength of various herbs.

Grass	Mean diameter of roots (mm)	Mean tensile strength (MPa)
Late Juncellus	0.38±0.43	24.50±4.2
Dallis grass	0.92±0.28	19.74±3.00
White Clover	0.91±0.11	24.64±3.36
VETIVER GRASS	*0.66±0.32*	*85.10±31.2*
Common Centipede grass	0.66±0.05	27.30±1.74
Bahia grass	0.73±0.07	19.23±3.59
Manila grass	0.77±0.67	17.55±2.85
Bermuda grass	0.99±0.17	13.45±2.18

Cheng et al (2003) supplemented Diti Hengchaovanich's root strength research by onducting further tests on other grasses. Table 5. Although vetiver has the second finest roots, its tensile strength is almost three times higher than all plants tested.

3.4 Hydraulic characteristics

When planted in rows, vetiver plants form thick hedges; their stiff stems allow these hedges to stand up at least 0.6-0.8m (2-2.6'), forming a living barrier to slow and spread runoff water. Properly planned, these hedges are very effective structures that spread and divert runoff water to stable areas or proper drains for safe disposal. Flume tests conducted at the University of Southern Queensland to study the design and incorporation of vetiver hedges into strip-cropping layout for flood mitigation confirmed the hydraulic characteristics of vetiver hedges under deep flows. Figure 4. The hedges successfully reduced flood velocity and limited soil movement; fallow strips suffered very little erosion, and a young sorghum crop was completely protected from flood damage (Dalton et al, 1996).

Figure 4: Hydraulic model of flooding through vetiver hedges

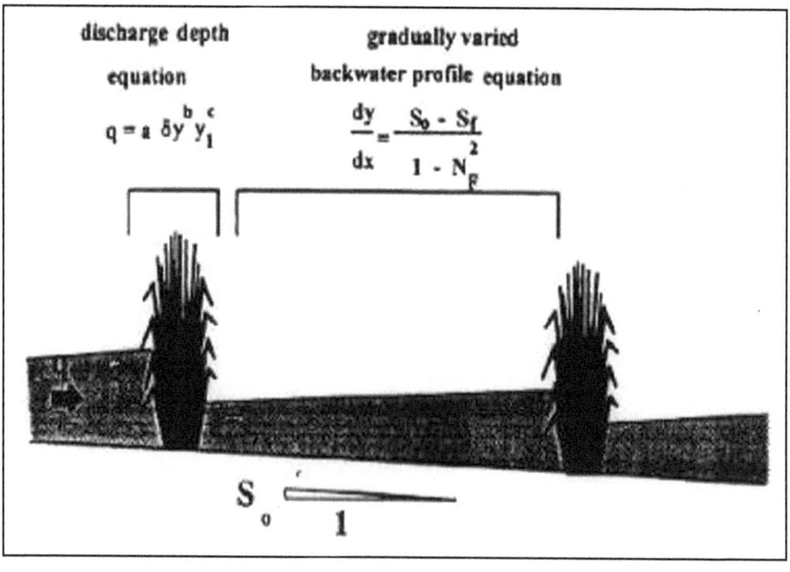

Where:
q = discharge per unit width
y = depth of flow y_1 = depth upstream S_o = land slope
S_f = energy slope N_F = the Froude number of flow

3.5 Pore water pressure

Vegetation cover on sloping lands increases water infiltration. Concerns have been raised that the extra water will increase pore water pressure in the soil and lead to slope instability. However, field observations actually show improvements. First, planted on contour lines or modified patterns of lines that trap and spread runoff water on the slope, vetiver's extensive root system and flow though effect distributes surplus water more evenly and gradually and helps prevent localized accumulation.

Second, the likely increase in infiltration is offset by a higher and gradual rate of soil water depletion by the grass. Research in soil moisture competition in crops in Australia (Dalton et al, 1996) shows that, under low rainfall conditions, this depletion would reduce soil moisture up to 1.5m (4.5') from the hedges. This increases water infiltration in that zone, leading to the reduction of runoff water and erosion rate. From a geotechnical perspective, these conditions help maintain slope stability. On steep (30-60°) slopes, the space between rows at 1m (3') VI (Vertical Interval) is very close. Therefore, moisture depletion would be greater and further improve the slope stabilisation process. However, to reduce this potentially harmful effect of vetiver on steep slopes in very high rainfall areas, as a precautionary measure, vetiver hedges could be planted on a gradient of about 0.5% as in graded contour terraces to divert the extra water to stable drainage outlets (Hengchaovanich, 1998).

3.6 Applications of VS for slope stabilization related to natural disaster mitigation and infrastructure protection

Given its unique characteristics, vetiver generally is very useful in controlling erosion on both cut and fill batters and on other slopes associated with road construction, and particularly effective in highly erodible and dispersible soils, such as sodic, alkaline, acidic and acid sulfate soils.

Vetiver planting has been very effective in erosion control or stabilisation in the following conditions:
- Slope stabilisation along highways and railways. Especially effective along mountainous rural roads, where the community

lacks sufficient funding for road slope stabilisation and where it often takes part in road construction.

- Dike and dam batter stabilisation, reduction of canal, riverbank and coastal erosion, and protection of hard structures themselves (e.g. rock riprap, concrete retaining walls, gabions, etc.).
- Slope above culvert inlets and outlets (culverts, abutments).
- Interface between cement and rock structures and erodible soil surfaces.
- As a filter strip to trap sediment at culvert inlets.
- To reduce energy at culvert outlets.
- To stabilize gully head erosion, when vetiver hedges are planted on contour lines above gully heads.
- To eliminate erosion caused by wave action, by planting a few rows of vetiver on the edge of the high water mark on big farm dam batters or river banks.
- In forest plantations, to stabilize the shoulders of access roads on very steep slopes as well as the gullies (logging paths/ways) that develop following harvests.

Given its unique characteristics, vetiver effectively controls water disasters such as flood, coastal and riverbank erosion, dam and dike erosion, and general instability. It also protects bridges, culvert abutments and interfaces between concrete/rock structures and soil. Vetiver is particularly effective in areas where the embankment fill is highly erodible and dispersible, such as sodic, alkaline, and acidic (including acid sulphate) soils.

3.7 Advantages and disadvantages of Vetiver System
Advantages:
- The major advantage of VS over conventional engineering measures is its low cost and longevity. For slope stabilisation in China, for example, savings are in the order of 85-90% (Xie, 1997 and Xia et al, 1999). In Australia, the cost advantage of VS over conventional engineering methods ranges from 64% to 72%, depending on the method used (Braken and Truong 2001). In summary, its maximum cost is only 30% of the cost of traditional measures. In addition annual maintenance costs are significantly reduced once vetiver hedgerows are established

- As with other bioengineering technologies, VS is a natural, environmentally-friendly way to control erosion control and stabilize land that 'softens' the harsh look of conventional rigid engineering measures such as concrete and rock structures. This is particularly important in urban and semi-rural areas where local communities decry the unsightly appearance of infrastructure development.
- Long-term maintenance costs are low. In contrast to conventional engineering structures, green technology improves as the vegetative cover matures. VS requires a planned maintenance program in the first two years; however, once established, it is virtually maintenance-free. Therefore, the use of vetiver is particularly well suited to remote areas where maintenance is costly and difficult.
- Vetiver is very effective in poor and highly erodible and dispersible soils.
- VS is particularly well suited to areas with low-cost labour forces.
- Vetiver hedges are a natural, soft bioengineering technique, an eco-friendly alternative to rigid or hard structures.

Disadvantages:
- The main disadvantage of VS applications is the vetiver's intolerance to shading, particularly within the establishment phase. Partial shading stunts its growth; significant shading can eliminate it in the long term by reducing its ability to compete with more shade-tolerant species. However, this weakness could be desirable in situations where initial stabilisation requires a pioneer to improve the ability of the micro-environment to host the voluntary or planned introduction of native endemic species.
- The Vetiver System is effective only when the plants are well established. Effective planning requires an initial establishment period of about 2-3 months in warm weather and 4-6 months in cooler times. This delay can be accommodated by planting early, and in the dry season.
- Vetiver hedges are fully effective only when plants form closed hedgerows. Gaps between clumps should be timely re-planted.

- It is difficult to plant and water vegetation on very high or steep slopes.
- Vetiver requires protection from livestock during its establishment phase.

Based on these considerations, the advantages of using VS as a bioengineering tool outweigh its disadvantages, particularly when vetiver is used as a pioneer species.

Worldwide evidence supports the use of VS to stabilize embankments. Vetiver has been used successfully to stabilize roadsides, amongst others, in Australia, Brazil, Central America, China, Ethiopia, Fiji, India, Italy, Madagascar, Malaysia, Philippines, South Africa, Sri Lanka, Venezuela, Vietnam, and the West Indies. Used in conjunction with geotechnical applications, vetiver has been used to stabilize embankments in Nepal and South Africa.

3.8 Combination with other types of remedy

Vetiver is effective both by itself and combined with other traditional methods. For example, on a given section of riverbank or dike, rock or concrete riprap can reinforce the underwater part, and vetiver can reinforce the top part. This tandem application creates a factor of stability and security (which are not always true and/or necessary). Vetiver can also be planted with bamboo, a plant traditionally used to protect riverbanks. Experience shows that using only bamboo has several drawbacks that can be overcome by adding vetiver. As noted previously washed out bamboo can create serious problems on rivers where there are low level bridge crossing.

3.9 Computer modelling

Software developed by Prati Amati, Srl (2006) in collaboration with the University of Milan determines the percentage or amount of shear strength that vetiver roots add to various soils under vetiver hedgerows. The software helps to assess vetiver's contribution to stabilize steep batters, particularly earthen levees. Under average soil and slope conditions, the installation of vetiver will increase slope stability by about 40%.

Using the software requires the operator to enter the following

geotechnical parameters related to a particular slope site:
- Soil type.
- Slope gradient.
- Maximum moisture content.
- Soil cohesion at a minimum.

The program provides the required number of plants per square meter and the distance between rows, considering the slope gradient. For example:
- a 30° slope requires six plants per square meter (i.e. 7-10 plants per lineal meter) and a distance between rows of about 1.7 m (5.7').
- a 45° slope requires 10 plants per square meter (i.e. 7-10 plants per lineal meter) and a distance between rows of about 1 m (3').

4. APPROPRIATE DESIGNS AND TECHNIQUES

4.1 Precautions
VS is a new technology. As a new technology, its principles must be studied and applied appropriately for best results. Failure to follow basic tenets will result in disappointment, or worse, adverse results. As a soil conservation technique and, more recently, a bioengineering tool, the effective application of VS requires an understanding of biology, soil science, hydraulics, hydrology, and geotechnical principles. Therefore, for medium to large-scale projects that involve significant engineering design and construction, VS is best implemented by experienced specialists rather than by local people themselves. However, knowledge of participatory approaches and community-based management are also very important. Thus, the technology should be designed and implemented by experts in vetiver application, associated with an agronomist and a geotechnical engineer, with assistance from local farmers.

Additionally, although it is a grass, vetiver acts more like a tree, given its extensive and deep root system. To add to the confusion, VS can exploit vetiver's different characteristics for different applications. For example, its deep roots stabilize land, its thick leaves spread water and trap sediment, and its extraordinary tolerance to hostile conditions allows it to rehabilitate soil and water contamination.

Failures of VS can, in most cases, be attributed to bad applications rather than the grass itself or the recommended technology. For example, in one case, vetiver was used in the Philippines to stabilize batters on a new highway. The results were very disappointing and failures resulted. It later surfaced that the engineers who specified the VS, the nursery that supplied the planting material, and the field supervisors and labourers who planted the vetiver, lacked previous experience or training in the use of VS for steep slopes stabilisation.

Experience in Vietnam shows that vetiver has been very successful employed when it is applied correctly. Not surprisingly, improper applications may fail. Applications in the Central Highlands of Vietnam show that vetiver has effectively protected road embankments. However, among mass applications on very high and steep slopes without benches along the Ho Chi Minh Highway, failures have resulted. In short, to ensure success, decision makers, designers and engineers who plan to use the Vetiver System for infrastructure protection should take the following precautions:

Technical precautions:
- To ensure success, the design should be created or checked by trained people.
- At least for the first few months while the plant is becoming established, the site should be internally stable against possible failure. Vetiver manifests its full abilities when mature, and slopes may fail during the intervening period.
- VS is applicable only to earthen slopes with gradients that should never exceed 45-50°
- Vetiver grows poorly in the shade, so planting it directly under a bridge or other shelter should be avoided.

Precautions for decision-making, planning and organisation:
- Timing: planning should consider the seasons and the time it takes to grow planting materials.
- Maintenance and repair: at an early stage, there is a period

during which vetiver is not yet effective. Planning and budgeting should anticipate replacement of some.

- Procurement: All inputs can and should be procured locally (labour, manure, planting materials, maintenance contracts). Employment opportunity provides an incentive for the local community to protect the plants during their infancy and adolescence, and to maintain the quality and sustainability of the works.
- Community involvement: As much as possible, local communities should be included in the design, materials procurement, and maintenance stages. Contracts with local people should be drafted, governing nurseries, quality/quantity specifications, and maintenance/protection.
- Timing: Decision makers should be ready to innovate and to consider VS in their planning and budgeting. For that, they need incentives to include such cost-effective methods in their plans, just as they have incentives - justified or not - to adopt more expensive conventional methods.
- Integration: Policy makers should recommend Vetiver System as part of a comprehensive approach to infrastructure protection, applied on a scale large enough to ensure a tangible increase in expertise and a gradual, spreading effect. VS should not be regarded merely as a fix for compromised local sites, despite its ability to provide a concise and immediate effect.

4.2 Planting time

The installation of vetiver plants is critical to the success and the cost of the project. Planting in dry season will require extensive and expensive watering. Experience in Central Vietnam shows that daily or twice daily watering is required to establish vetiver in the extremely harsh conditions in sand dunes. Growth is stunted in the absence of watering. Since it is difficult to select the best time to plant masses of plant material on cut slopes along the Ho Chi Minh Highway, for example, mechanical watering is required daily for the first few months.

Vetiver generally needs 3-4 months to become established, sometimes up to 5-6 months under adverse conditions. Since vetiver is fully effective at the age of 9-10 months, mass plantings should occur at the beginning of the rainy season (i.e. nursery development and production of plant material should be planned to meet that mass planting schedule).

Particularly in North Vietnam, it is possible to plant during the winter-spring period. When temperatures descend lower than 10°C (50°F) in North Vietnam, the grass does not grow. However, it can survive the cold weather and resumes growing immediately when the winter rain starts and the weather warms.

In central Vietnam, where air temperature usually stays above 15°C (59°F), mass planting occurs at the beginning of spring. Nurseries will require more care to ensure good growth and multiplication of the slips.

4.3 Nursery
The success of any project depends on good quality and sufficient numbers of vetiver slips. Large nurseries generally are not required to provide sufficient plant material. Instead, individual farmer households can set up and supervise small nurseries (a few hundred square meters each). They will be contracted and paid by the project according to the number of slips they can provide upon request.

4.4 Preparation for vetiver planting
In cases where mass planting of vetiver involves the participation of local people, an effective planting campaign should include the following steps:

Step 1:   Experts visit the sites, and conduct a survey to iden-
           tify problems and design the application of the tech-
           nology;
Step 2:   Discuss the problems and alternative solutions
           with local people;
Step 3:   Use workshops and training courses to introduce
           the new technology;
Step 4:   Organize the trial implementation, by establishing

	nurseries, contracting to purchase plant material, maintenance, etc.;
Step 5:	Monitor the implementation;
Step 6:	Discuss results of the pilot, following workshop, field exchange visit, etc.;
Step 7:	Organize mass planting.

In cases where specialized companies undertake the mass planting, steps 1, 4, 5 are recommended. However, local participation is still advisable to raise awareness, avoid vandalism, and ensure that the slips are protected from animals.

4.5 Layout specifications
4.5.1 'Upland' natural slope, cut slope, road batter, etc.
To stabilize upland natural slopes, cut slopes, and road batters, the following specifications may apply:

- Bank slope should not exceed 1(H) [horizontal]:1(V) [vertical] or 45°, gradient of 1.5:1 is recommended. Shallower gradients are recommended wherever possible, especially on erodible soils and/or in high rainfall areas.
- Vetiver should be planted across the slope on approximate contour lines with a Vertical Interval (VI) between 1.0-2.0m (3-6') apart, measured down the slope. Spacing of 1.0m (3') should be used on highly erodible soil, which can increase up to 1.5-2.0m (4.5-6') on more stable soil.
- The first row should be planted on the top edge of the batter. This row shall be planted on all batters that are taller than 1.5m (4.5').
- The bottom row should be planted at the bottom of the batter at the toe of the slope and on cut batter along the edge of table drain.
- Between these rows, vetiver should be planted as specified above.
- Benching or terracing 1-3 m (3-9') in width for every 5-8 m (15-24') VI is recommended for slopes that are taller than 10 m (30').

4.5.2 *Riverbanks, coastal erosion, and unstable water retaining structures*

For flood mitigation and coastal, riverbank and dike/embankment protection, the following layout specifications are recommended:

- Maximum bank slope should not exceed 1.5(H):1(V). Recommended bank slope is 2.5:1. Note: the sea dike system in Hai Hau (Nam Dinh) is built with bank slope of 3:1 to 4:1.
- Vetiver should be planted in two directions:
  - For bank stabilisation, vetiver should be planted in rows parallel to flow direction (horizontal), on approximate contour lines 0.8-1.0m (2.5-3') apart (measured down slope). A recent layout specification to protect the sea dike system in Hai Hau (Nam Dinh) included spacing between rows lowered to 0.25 m. (.8').
  - To reduce flow velocity, vetiver should be planted in rows normal (right angle) to the flow at spacing between rows of 2.0m (6') for erodible soil and 4.0m (12') for stable soil. As added protection, normal rows are planted 1.0m (3') apart on the river dike in Quang Ngai.
- The first horizontal row should be planted at the crest of the bank and the last row should be planted at the low water mark of the bank. Note: since the water level at some locations changes seasonally, vetiver can be planted much further down the bank when the time is right.
- Vetiver should be planted on the contour along the length of the bank between the top and bottom rows at the spacing specified above.
- Due to high water levels, bottom rows may establish more slowly than upper rows. In such cases, the lower rows should be planted when the soil is driest. Some VS applications protect anti-salinity dikes; in those cases, the water may become more saline at certain times of the year, which may affect the growth of vetiver. Experiences in Quang Ngai show that vetiver can be replaced by some local salt-tolerant varieties, including the mangrove fern.
- For all applications, VS can be used in combination with other traditional, structural measures such as rock or concrete riprap, and retaining walls. For example, the lower part of the

dike/embankment can be covered by the combination of rock riprap and geo-textile while the upper half is protected with vetiver hedgerows.

4.6 Planting specifications

- Dig trenches that are about 15-20cm (6-8") deep and wide.
- Place well-rooted plants (with 2-3 tillers apiece) in the centre of each row at 100-120mm (4-5") intervals for erodible soils, and at 150mm (6") for normal soils.
- Since soil on slopes, road batters and filled dike/embankment is not fertile, it is recommended that potted or tube stock be used for large scale mass planting and rapid establishment. Adding a bit of good soil-manure mixture (slurry) is even better. To protect natural river banks where the soil is usually fertile and initial watering can be ensured without extra effort, bare root planting is sufficient.
- Cover roots with 200-300mm (8-12") of soil and compact firmly.
- Fertilize with Nitrogen and Phosphorus such as DAP (Di -Ammonium Phosphate) or NPK (note from experience vetiver does not respond significantly from potash applications) at 100g (3.5oz) per linear meter (row). The same amount of lime may be necessary when planting in acid and sulfate soil.
- Water within the day of planting.
- To reduce weed growth during the establishment phase, a pre-emergent herbicide such as Atrazine may be used.

4.7 Maintenance

Watering

- In dry weather, water every day during the first two weeks after planting and then every second day.
- Water twice weekly until the plants are well established.
- Mature plants require no further watering.

Replanting

- During the first month after planting, replace all plants that fail to establish or wash away.
- Continue inspections until the plants are suitably established.

Weed control

- Control weeds, especially vines, during the first year.
- DO NOT USE RoundUp (glyphosate) herbicide. Vetiver is very sensitive to glyphosate, so it should not be used to control weeds between rows.

Fertilizing
On infertile soil, DAP or NPK fertilizer should be applied at the beginning of the second wet season.

Cutting
After five months, regular cutting (trimming) is also very important. Hedgerows should be cut down to 15-20 cm (6-8") above the ground. This simple technique promotes the growth of new tillers from the base and reduces the volume of dry leaves that otherwise can overshadow young slips. Trimming also improves the appearance of dry hedgerows and may minimize the danger of fire.

Fresh cut leaves can also be used as cattle fodder, for handicraft, and even roof thatch. Please note that vetiver planted for the purpose of reducing natural disasters should not be overused for secondary purposes.

Subsequent cuttings can be done two or three times a year. Care should be taken to ensure the grass has long leaves during the typhoon season. Vetiver can be cut immediately after the typhoon season ends. Another suitable cutting time could be around 3 months before the typhoon season begins.

Fencing and caring
During the several-month establishment period, fencing and care may be required to protect vetiver from vandalism and cattle. The old stems of mature vetiver are tough enough to discourage cattle. Where necessary, it is advisable to fence the area to protect the grass during the first few months after planting.

5. VETIVER SYSTEMS APPLICATIONS FOR NATURAL DISASTER REDUCTION AND INFRASTRUCTURE PROTECTION IN VIETNAM

5.1 VS application for sand dune protection in Central Vietnam

A vast area, more than 70,000 ha (175,000 acres), along the coastline of Central Vietnam is covered by sand dunes where the climatic and soil conditions are very severe. Sand blast often occurs as sand dunes migrate under the action of wind. Sand flow also takes place frequently due to the action of numerous permanent and temporary streams. Blown sand and sand flow transport huge amounts of sand from dunes landward onto the narrow coastal plain. Along the Central Vietnam coastline, giant sand "tongues" bite into the plain day after day. The Government has long implemented a forestation program using such varieties as Casuarinas, wild pineapple, eucalyptus, and acacia. However, when fully and well established, they may help reduce only blown sand. Until now, there has been no way to reduce sand flow (trees can not stabilize sand dunes, especially on their 'slip-face', this was tried in North Africa by FAO at great expense and failed).

In February 2002, with financial support from the Dutch Embassy Small Program and technical support from Elise Pinners and Pham Hong Duc Phuoc, Tran Tan Van from RIGMR initiated an experiment to stabilize sand dunes along the Central Vietnam coastline. A sand dune was badly eroded by a stream that served as a natural boundary between farmers and a forestry enterprise. The erosion occurred over several years, resulting in a mounting conflict between the two groups. Vetiver was planted in rows along the contour lines of the sand dune. After four months it formed closed hedgerows and stabilized the sand dune. The forestry enterprise was so impressed that it decided to mass plant the grass in other sand dunes and even to protect a bridge abutment. Vetiver further surprised local people by surviving the coldest winter in 10 years, when the temperature descended lower 10°C (50°F), forcing the farmers to twice replant their paddy rice and Casuarinas. After two years, the local species (primarily Casuarinas and wild pineapple) became re-established. The grass itself faded away under the shade of these trees, having accomplished its mission.

The project proved again that, with proper care, vetiver could survive very hostile soil and climatic conditions - photo 2.

According to Henk Jan Verhagen from Delft University of Technology (pers. comm.), vetiver may be equally effective in reducing blown sand (sand drift). For this purpose, the grass could be planted across the wind direction, especially at low places between sand dunes, where the wind velocity typically increases. On China's Pintang Island, off the coast of Fujian Province, vetiver hedges effectively reduced wind velocity and blow sand.

Following the success of this pilot project, a workshop was organized in early 2003. More than 40 representatives from local government departments, different NGOs, the University of Central Vietnam, and coastal provinces participated. The workshop helped the authors of this book and other participants to compile and synthesize local practices, particularly regarding planting times, watering, and fertilizing. Following the event, World Vision Vietnam decided in 2003 to fund another project in the Vinh Linh and Trieu Phong districts in Quang Tri province to employ vetiver for sand dune stabilisation - photos 3-7.

5.1.1 Trial application and promotion of VS for sand dune protection in coastal province of Quang Binh

Photo 2: Sand flow in Le Thuy (Quang Binh) in 1999: the foundation of a pumping station (upper); this woman's three-room brick house is collapsing because sand has been blown from foundation (lower).

**Photo 3: Upper: site overview; lower: early April 2002,
one month after planting.**

Photo 4: Upper: early July 2002, four months after planting; lower:
November 2002, dense rows of grass have been established.

Photo 5: Upper: Vetiver nursery; lower: November 2002, mass planting.

Photo 6: Upper: Vetiver protects bridge abutment along National Highway nr.1; lower: December 2004, local species have replaced vetiver.

Photo 7: Upper: mid-February 2003, post-workshop field trip; Note: Vetiver survives even the coldest winter in 10 years; lower: June 2003, farmers from Quang Tri province visit a local nursery during a World Vision Vietnam-sponsored field trip.

5.2 VS application to control river bank erosion

5.2.1 VS application for river bank erosion control in Central Vietnam

Within the framework of the same Dutch Embassy project mentioned above, vetiver was planted to halt erosion on a riverbank, on the bank of a shrimp pond, and on a road embankment in Da Nang City. In October 2002, the local Dike Department also mass planted the grass on bank sections of several rivers. Thereafter, the city authority decided to fund a project on cut slope stabilisation by installing vetiver along the mountainous road leading to the Banana project in Da Nang, illustrating the pace of adoption - photos 8-10.

Photo 8: Upper: December 2004: Vetiver, combined with rock riprap, flourishes after two flood seasons (Da Nang); lower: planted by local farmers, vetiver protects their shrimp ponds.

Photo 9: Upper: March 2002: VS trial at the edge of a shrimp pond, where a canal drains flood water to Vinh Dien River; lower: November 2002: mass planting combined with rock riprap to protect bank along Vinh Dien river.

Photo 10: Upper: Vetiver and rock riprap and concrete frame protect an embankment; lower: a bend on Perfume River bank in Hue - protected with vetiver.

5.2.2 VS trial and promotion for river bank protection in Quang Ngai

As another result of this pilot project, vetiver was recommended for use in another natural disaster reduction project in Quang Ngai province, funded by AusAid. With technical support by Tran Tan Van in July 2003, Vo Thanh Thuy and his co-workers from the provincial Agricultural Extension Centre.

Photo 11: Vetiver grass planted on river dike along Tra Bong River (upper) and lining the sides of an anti-salinity estuary dike along the same river (lower).

60

Photo 12: Upper: severely eroded bank of the Tra Khuc River, at Binh Thoi Commune; lower: primitive sand bag protection.

Photo 13: Upper: Community members plant vetiver; lower: November 2005: bank remains intact following the flood season.

planted the grass at four locations, irrigation canals in several districts and a seawater intrusion protection dikes. Vetiver thrived in all locations and, despite its young age, survived a flood in the same year - photos 11-13.

Following these successful trials, the project decided to mass plant vetiver on other dike sections in three other districts, in combination with rock riprap. Design modifications introduced to better adapt vetiver to local conditions include planting mangrove fern and other salt-tolerant grasses on the lowest row to better withstand high salinity and to effectively protect the embankment toe. Encouragingly, local communities are more readily using vetiver to protect their own lands

5.2.3 VS application to control river bank erosion in the Mekong Delta

With William Donner Foundation financial support and Paul Truong's technical help, Le Viet Dung and his colleagues at Can Tho University initiated riverbank erosion control projects in the Mekong Delta. The area experiences long periods of inundation (up to five months) during the flood season, with significant difference in water levels, up to 5 m (15'), between dry and flood seasons, and powerful water flow during flood season. Further, the riverbanks consist of soils ranging from alluvial silt to loam, which are highly erodible when wet. Due to the improved economy of recent years, most boats travelling on rivers and canals are motorized, many with powerful engines that aggravate riverbank erosion by generating strong waves. Nevertheless, vetiver stands its ground, protecting large areas of valuable farm land from erosion - photos 14 and 15.

A comprehensive vetiver program has been established in An Giang Province, where annual floods reach depths of 6 m (18'). The province's long, 4932 km (3065 miles), canal system requires annual maintenance and repair. A network of dikes, 4600 km long, protects 209,957 ha (525,000 acres) of prime farmland from flood. Erosion on these dikes is about 3.75 Mm3/year and required USD 1.3 M to repair.

The area also includes 181 resettlement clusters, communities built on dredged materials that also require erosion control and protection from flooding. Depending on the locations and flood depth, vetiver has been used successfully alone, and together with other vegetation to stabilize these areas. As a result, vetiver now lines rigorous sea and river dike systems as well as riverbanks and canals in the Mekong Delta. Nearly two million polybags of vetiver, a total of 61 lineal km (38 miles), were installed to protect the dikes between 2002 and 2005

- photos 14 -15.

Between 2006 and 2010, the 11 districts of An Giang province are expected to plant 2025 km (1258 miles) of vetiver hedges on 3100 ha (7660 acres) of dike surface. Left unprotected, 3750 Mm3 of soil likely will be eroded and 5 Mm3 will have to be dredged from the canals. Based on 2006 current costs, total maintenance costs over this period would exceed US $ 15.5 M in this province alone. Applying the Vetiver System in this rural area will provide extra income to the local people: men to plant, and women and children to prepare polybags.

Photo 14: In An Giang vetiver stabilizes a river dike (upper), and a natural river bank (lower).

64

Photo 15: Upper: Vetiver borders the edge of flood resettlement centres; lower: the red markers delineate about 5 m (15') of dry land saved by vetiver.

5.2.4 Vetiver System application to control severe river bank erosion in Cambodia

The water level of the Mekong river section in Cambodian fluctuates widely, reaching 15m and more during the flood season above the level during the dry season. The combination of very fast current and wave action during the annual rainy and flood season causes severe bank erosion, averaging between 5-10m every year. General soil erosion on the alluvial plain is between 10 and 30cm each year. The loss of this fertile alluvial plain severely affects agriculture production, valuable urban land and infrastructure stability along the river. In addition the water is very muddy and high in sediment load.

A project was initiated in 2006 to stabilise a 200m long stretch of the Mekong bank north of Phnom Penh, the capital of Cambodia, which has been severely eroded and will eventually wash away the national highway to northern regions. On this site, severe erosion occurs every year and after 10 years, 50m have disappeared that translates into a loss of 50 m x 200 m (width) = 10,000 m2 or 1 hectare! Various stabilising options such as conventional hard structures including gabion and rock wall and local vegetation, bamboo, were considered, but these measures are either ineffective and/or too expensive to implement. As a result, Vetiver System technology was implemented as a last resort. (Tuon Van, Coordinator Cambodian Vetiver Network, pers.com.)

Results of riverbank stabilisation works in Australia, China, Madagascar and Vietnam have shown that the stiff vetiver shoots reduced flow velocity, hence erosive power, and its deep and extensive root system reinforce the soil and holds it firmly to the ground, resulting in a very effective stabilising mechanism. It is expected that this mechanism is also effective whether the vetiver is alive or dead in the short term. Therefore when fully established vetiver would control/reduce the erosion on the bank of the Mekong under flood.

South African experience and Chinese research showed that vetiver could survive up to 3 months under clean, clear water and still conditions. However it is not known how long vetiver could survive

Photo 16: Eroded bank before (upper) and after earthwork (lower).

under muddy, turbulent and fast flowing conditions. It was expected that muddy river water would affect vetiver growth due to low light transmission.

The eroded bank was first reshaped and firmly packed to 300 gradient, with a vertical drop of approximately 8m - photo 16. To provide maximum protection vetiver was planted in a grid pattern, both on the contour line along the bank (horizontal row) to reduce wave erosion and up and down the slope (vertical row) to reduce flow velocity. The spacing of the horizontal rows is 1m apart with plant density of 10 plant/m and vertical rows are spaced at 2m apart with plant density of 5plant/m. The planting was fertilized with both manure and chemical fertilizers to ensure maximum growth - photo 17.

Photo 17: One month (upper) and seven months after planting (lower).

Seven months after planting, with intensive maintenance (fertiliser, watering and weed control), the vetiver stand was 1.5m high. To test the survival rate of vetiver under muddy conditions, some sections of the slope were trimmed down to 50cm high before flooding.

Photo 18: Flood water started coming up (upper); and dead looking vetiver on lower slope after water retreats (lower).

As expected the site was fully flooded nine months after planting, covering the whole slope and higher ground. Although vetiver was not fully mature it successfully stopped the erosion. On the upper part of the slope, where submergence time was shorter (up to two months) and shallower depth, vetiver growth was not affected and continued to grow under water. It was expected that plants on the bottom part of the slope, which was submerged for 6 months and under 14m of muddy water, vetiver would be badly affected. But surprisingly, although they all look dead only a few were actually dead - photo 18.

The followings observations and conclusion were recorded:
- the first 3 rows at the base of the slope all survived! These rows were submerged for 5-6 months under 14m of muddy water. They were not cut (1.5m tall) before the water started to rise.
- the next 5 rows up, the vetiver was cut to 50cm and they all died because they were all covered by mud.
- the rest of the slope starting from rows 9, were also cut to 50cm, but all survived because the mud didn't fully cover them.
- plants from upper section actually grew under water during the flood.
- Uncut plants survived better - photo 19 lower.
- and the thick mud cover killed them - photo 19 upper.

But most importantly, vetiver planting did not only stop the bank erosion, it also accumulated a thick cover of alluvial silt between the rows - photo 20. On closer examination, this thick silt cover was the cause of vetiver death in some sections of the slope. Where the mud cover was not too thick, vetiver shoots emerged later - photo 21. The dead plants have been replaced and planting will be extended to even lower area of the slope - photo 21.

The above results show that:
- vetiver can survive up to 5-6 months under 14m of muddy water
- uncut shoots improve its survival under water
- vetiver is killed when covered with or buried under thick alluvial mud

70

Photo 19: Dead vetiver due to thick alluvial silt deposit (upper); and re-growth if the mud was not too thick (lower).

It can be concluded that when correctly designed and implemented, vetiver planting is very effective in controlling erosion on the bank of fast flowing river even under flood conditions and deep and prolonged submergence in muddy water. In addition it encourages alluvial silt deposition and over time may eventually reclaim the eroded banks.

**Photo 20: Upper and lower: alluvial silt deposit between vetiver rows on
lower part of the slope.**

Photo 21: Fully recovered after the flood (upper) and new planting on the bare lower section of slope (lower).

5.3 VS application for coastal erosion control

Huge sea dikes with revetment protection built from traditional "hard" material such as block concretes or big rock have given good results. The height of these sea dikes should be sufficient to protect the area inside the coastal flood defence system. However they are quite costly to implement and materials are not always available. In order to reduce total cost, the traditional revetment could be replaced by cheaper materials. A combination of "hard" and "soft" materials is a good alternative solution. Vetiver grass is well-known as bioengineering species in stabilizing inner slopes, reducing run-off and controlling soil loss. Recently, it has been planted on outer slope as sea dike protection as well - photo 22. However the understanding of the processes and properties between waves and Vetiver grass is still limited.

Photo 22: Vetiver planted on the inside of Hai Hau sea dike (upper) and outside (lower).

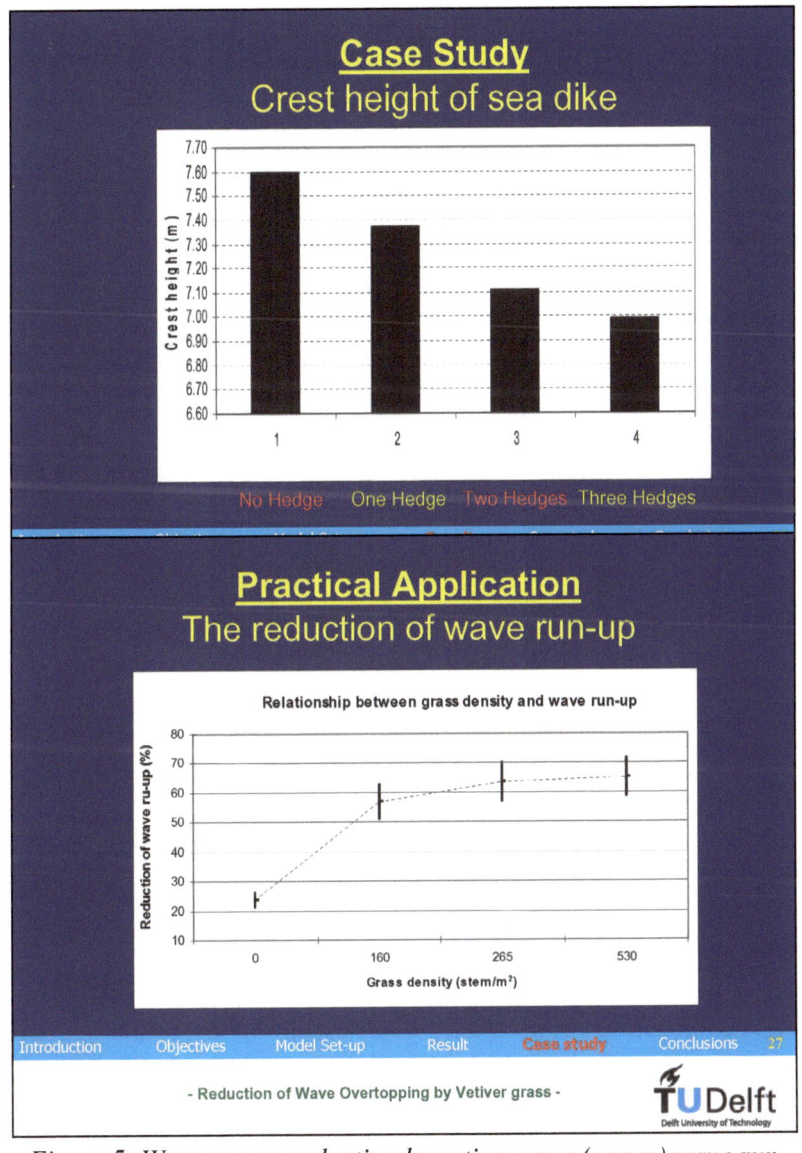

Figure 5: Wave run up reduction by vetiver rows (upper) wave run up reduction by plant density (lower).

Recently the Department of Hydraulic Engineering at Delft Technical University in Holland conducted research on the use of vetiver grass on the dike outer slope to reduce wave run up (overtopping discharges)

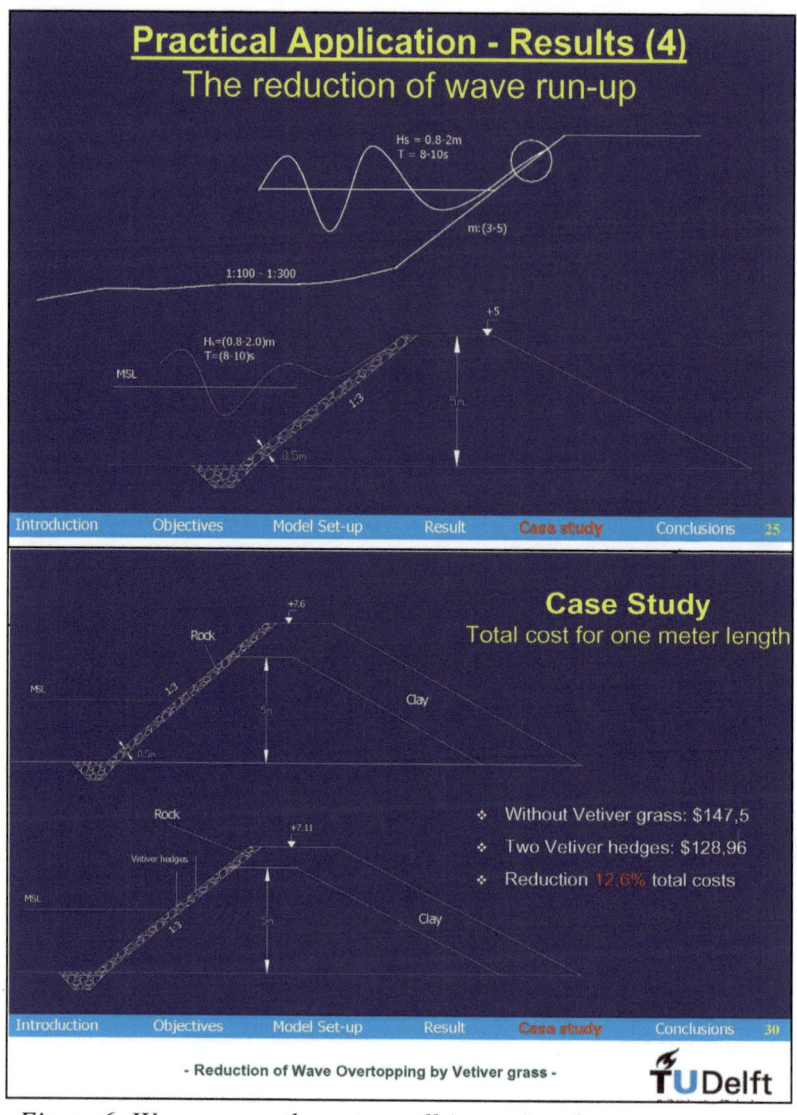

Figure 6: Wave run on the outer wall (upper) and construction cost saving with and without vetiver (lower).

so that sea dike crest can be reduced. A physical model was conducted using fully grown Vetiver grass hedges and wave parameters in front of the hedges. Experimental results have shown that:

- Resistance to flow by Vetiver hedge varies with grass density

- figure 5.
- Resistance (Manning coefficient) varies with flow depth and vetiver provided 2.5 times greater resistance than bare slope
- Vetiver grass hedges can withstand flow of backwater up to nearly 0.4m depth.
- The roughness coefficient of Vetiver grass, depending on grass density varies between 0.33 and 0.41.
- The reduction of wave overtopping of more than 60% - figure 5. When this model was applied on a sea dike in Vietnam, result shows that a reduction of 0.5m of the crest height is feasible for upgrading the present sea dike in Hai Hau, Vietnam. In dollar term, the cost of construction per meter length of $147.5 without vetiver, reduces to $128.96 when two vetiver rows are planted on the outer slope, a reduction of 12.6%.

This case shows that Vetiver grass is a good solution for sea dikes in order to reduce wave run-up on the outer slope and decreases the cost for sea dike upgrading - figure 6 (Vu Minh Anh, 2007).

With support of the William Donner Foundation and with technical support by Paul Truong, Le Van Du from Ho Chi Minh City Agro-Forestry University in 2001 initiated work on acid sulfate soil to stabilize canal and irrigation channels and the sea dike system in Go Cong province. Vetiver grew vigorously on the embankments in just a few months, despite poor soil. It is now protecting the sea dike, preventing surface erosion, and facilitating the establishment of endemic species - photo 23 and 24.

Photo 23: Planted behind natural mangrove on an acid sulfate soil sea dike in Go Cong province, vetiver reduces surface erosion and fosters the re-establishment of local grasses.

Photo 24: In North Vietnam; Upper: Vetiver planted on outer side of a newly built sea dike in Nam Dinh province; lower: on the inner side of the dike, planted by the local Dike Department.

5.4 VS application to stabilize road batter

Following successful trials by Pham Hong Duc Phuoc (Ho Chi Minh City Agro-Forestry University) and Thien Sinh Co. in using vetiver to stabilize cut slopes in Central Vietnam, in 2003 the Ministry of Transport authorized the wide use of vetiver to stabilize slopes along hundreds of kilometres of the newly constructed Ho Chi Minh Highway and other national, provincial roads in Quang Ninh, Da Nang, and Khanh Hoa provinces - photo 25.

Photo 25: Upper: Vetiver stabilizes cut slopes along the Ho Chi Minh Highway; lower: both alone and in combination with traditional measures.

This project is certainly one of the largest VS applications in infrastructure protection in the world. The entire Ho Chi Minh

Highway is more than 3000 km (1864 miles) long. It is being and will be protected by vetiver planted under a variety of soils and climate

Photo 26: Upper - If not properly protected rock/soil from this waste dump will wash far downstream. Lower - impacting a downstream village in A Luoi district, Thua Tien Hue province.

from skeletal mountainous soils and cold winter in the North to extremely acidic acid sulphate soil and hot, humid climate in the South. The extensive use of vetiver to stabilize cut slopes works, for example:

- Applied primarily as a slope surface protection measure, it greatly reduces run-off induced erosion, that would otherwise wreak havoc downstream - photo 26.
- By preventing shallow failures, it stabilizes cut slopes which greatly reduces the number of deep slope failures - photo 27.
- In some cases where deep slope failures do occur, vetiver still

81

does a very good job in slowing down the failures and reducing the failed mass, and;
* It maintains the rural aesthetic and eco-friendliness of the road.

Photo 27: Da Deo Pass, Quang Binh: Upper: Vegetation cover is destroyed, revealing ugly and continuous failures of cut slopes; lower: Vetiver rows on top of the slope very slowly squeeze down, considerably reducing the failed mass.

On a road leading to the Ho Chi Minh Highway Pham Hong Duc Phuoc demonstrated clearly how VS should be applied, as well as its effectiveness and sustainability - photo 28.

He carefully monitored the development of vetiver: its establishment (65-100%), togrowth (95-160 cm (37-63")) after six months), tillering rate (18-30 tillers per plant), and root depth on the batter - table 6.

Table 6: Vetiver root depth on Hon Ba road batters.

	Position on the batter	Root depth (cm/inch)			
		6 months	12 months	1.5 year	2 years
	Cut Batter				
1	Bottom	70/28	120/47	120/47	120/47
2	Middle	72/28	110/43	100/39	145/57
3	Top	72/28	105/41	105/41	187/74
	Fill Batter				
4	Bottom	82/32	95/37	95/37	180/71
5	Middle	85/33	115/45	115/45	180/71
6	Top	68/27	70/28	75/30	130/51

The successes and failures using vetiver to protect cut slopes along the Ho Chi Minh Highway are instructive:

- Slopes must first be internally stable. Since vetiver is most helpful at maturity, slopes may fail in the interim. Vetiver begins to stabilize a slope at three to four months, at earliest. Therefore, the timing of planting is also very important if slope failure during the rainy season is to be avoided.
- Appropriate slope angle should not exceed 45-50°.
- Regular trimming will ensure continued growth and tillering of the grass, and thus ensure dense, effective hedgerows.

Photo 28: Pham Hong Duc Phuoc, a road protection project in Khanh Hoa province, road to Hon Ba): left two photos: severe erosion on newly built batter occurs after only a few rains; right two photos: eight months after vetiver planting: Vetiver stabilized this slope, totally stopping and preventing further erosion during the next wet season.

6. CONCLUSIONS

Following considerable research and the successes of the many applications presented in this handbook, we now have enough evidence that vetiver, with its many advantages and very few disadvantages, is a very effective, economical, community-based and environmentally-friendly sustainable bioengineering tool that protects infrastructure and mitigates natural disasters, and, once established, the vetiver plantings will last for decades with little, if any maintenance. VS has been used successfully in many countries in the world, including Australia, Brazil, Central America, China, Ethiopia, India, Italy, Malaysia, Nepal, Philippines, South Africa, Sri Lanka, Thailand, Venezuela, and

Vietnam. However, it must be stressed that the most important key to success are good quality planting material, proper design, correct planting techniques .

7. REFERENCES

Bracken, N. and Truong, P.N. (2 000). Application of Vetiver Grass Technology in the stabilization of road infrastructure in the wet tropical region of Australia. Proc. Second International Vetiver Conf. Thailand, January 2000.

Cheng Hong, Xiaojie Yang, Aiping Liu, Hengsheng Fu, Ming Wan (2003). A Study on the Performance and Mechanism of Soil-reinforcement by Herb Root System. Proc. Third International Vetiver Conf. China, October 2003.

Dalton, P. A., Smith, R. J. and Truong, P. N. V. (1996). Vetiver grass hedges for erosion control on a cropped floodplain, hedge hydraulics. Agric. Water Management: 31(1, 2) pp 91-104.

Hengchaovanich, D. (1998). Vetiver grass for slope stabilization and erosion control, with particular reference to engineering applications. Technical Bulletin No. 1998/2. Pacific Rim Vetiver Network. Office of the Royal Development Project Board, Bangkok, Thailand.

Hengchaovanich, D. and Nilaweera, N. S. (1996). An assessment of strength properties of vetiver grass roots in relation to slope stabilisation. Proc. First International Vetiver Conf. Thailand pp. 153-8.

Jaspers-Focks, D.J and A. Algera (2006). Vetiver Grass for River Bank Protection. Proc. Fourth Vetiver International Conf. Venezuela, October 2006.

Le Van Du, and Truong, P. (2003). Vetiver System for Erosion Control on Drainage and Irrigation Channels on Severe Acid Sulphate Soil in Southern Vietnam. Proc. Third International Vetiver Conf. China, October 2003.

Prati Amati, Srl (2006). Shear strength model. "PRATI ARMATI Srl" info@pratiarmati.it .

Truong, P. N. (1998). Vetiver Grass Technology as a bio-engineering tool for infrastructure protection. Proceedings North Region Symposium. Queensland Department of Main Roads, Cairns

August, 1998.

Truong, P., Gordon, I. and Baker, D. (1996). Tolerance of vetiver grass to some adverse soil conditions. Proc. First International Vetiver Conf. Thailand, October 2003.

Xia, H. P. Ao, H. X. Liu, S. Z. and He, D. Q. (1999). Application of the vetiver grass bio-engineering technology for the prevention of highway slippage in southern China. International Vetiver Workshop, Fuzhou, China, October 1997.

Xie, F.X. (1997). Vetiver for highway stabilization in Jian Yang County: Demonstration and Extension. Proceedings abstracts. International Vetiver Workshop, Fuzhou, China, October 1997.

INDEX